Plumber 'Speaks'

Author **Sherman Turner**

Plumber
'Speaks'

Copyright © 2021 Sherman Turner
All rights reserved. No part of this document may be reproduced or transmitted in any form or by any means, electronic, mechanical, photocopying, recording, or otherwise, without prior written permission of and by Sherman Turner.

The Table of Contents

DEDICATION iiii
Book Introduction v
Plumbing .... **viii**
  Chapter #1 Plumbing 1
  Chapter #2 Lavs & Vanity Sinks 6
  Chapter #3 Kitchen Sinks 10
  Chapter #4 Toilets 14
  Chapter #5 Water 24
  Chapter #6 Hot Water Tanks 28
  Chapter #7 Sewers and Drains 36
  Chapter #8 Sump Pumps 42
  Chapter #9 Plumbing Answers 45
  Chapter #10 Water, PEX - Piping 59
Cost Estimating **74**
  Chapter #11 Cost Estimating 75
About the Author **93**

Plumber 'Speaks'

DEDICATION

This book is dedicated to *Rachel Turner* my wife, my daughter *Sabrinna Turner*, my boys *Michael*, *Nelson*, *Timothy* and Jimmy.

The *Plumbers Union Local 22*, of Buffalo, NY. And, very special thank you to *Willie Roberson, Shaun Roberson, Charles Roberson* and *Earl Roberson,* for taking really good care of me.

Book Introduction

The Author and Master Plumber, *Sherman Turner* realize that in today's tough economy many people are wanting to know how to upkeep their homes and repairs at the same time while trying to increase savings.

Therefore, *Plumber 'Speaks'* book will enlighten your efforts towards saving $100 and $1K the easy way! The information in this **book** is designed to help make your cost saving decisions more lucrative.

Master Plumber, Sherman Turner has work as a Plumbing Inspector, a Journeyman Plumber, and a Plumbing Estimator. As a businessman he uses all three (3) levels of his experience to serve the people in the communities.

It is a known fact that most plumbers depend on your ignorance as their measurement for writing up your bill. At my three (3) levels of plumbing experience I found out this is true. That is why I decided to try to write a book that will benefit the consumer's rather than the Plumbers.

This book purpose is not to help you become a do-it-yourselfer (**DIY**) book but more so as a source of information that is "not privy" too many. As a retired Master Plumber and small businessperson this information, will benefit small businesses and all homeowners also.

Many **homeowners** and **small businesses** are getting ripped-off by unscrupulous large **Contractors**. This book will help you be better informed to increase your home savings.

This book by Master Plumber specializing estimating will give new informational plumbing charts, so you will be better informed. Our detailed answers to your many questions, is "priceless." Plus, the charts help **small businesses** become more competitive to win more jobs.

Plumber 'Speaks' is the first **book** of its kind with many dynamic answers helping to reduce plumbing costs and Contractor rip-offs. These new dynamic answers benefit the **consumers, homeowners** and **contractors.**

Plumbing

Chapter #1

Plumbing

Plumbing originated during ancient civilizations as they developed public baths and needed to provide potable water and wastewater removal for larger numbers of people.

Standardized plumbing pipes were made to prevent leaks in the settlements of Civilization. During early BC.

The word *"Plumber"* dates from the *"Roman Empire days."* the waste disposal system consisted of collecting waste and dumping it on the ground or into a river.

"Plumbing System" means the water supply and distribution pipes; Plumbing Fixtures and traps; soil, waste, and vent pipes; sanitary and storm sewers and building drains, including their respective connections, devices and within the premises for water-treating the equipment. Add now Nuclear Facilities piping and systems and Chemical Plants with acid systems.

Plumber 'Speaks'

What is the purpose of plumbing?

Plumbing uses pipes, valves, **Plumbing** Fixtures, Tanks, and other apparatuses to convey fluids. Heating and Cooling (HVAC), waste removal, and potable water delivery are among the most common uses for Plumbing, but it is not limited any of these

What are the different types of **Plumbing Systems**?

The **Plumbing** in any building serves two main purposes. The first is to bring water into the structure for human use, and the second is to remove wastewater of various types. There are **three** (3) main types of *"Plumbing Systems"* potable water, sanitary drainage and stormwater drainage.

What is *Plumbing* and its importance?

Plumbing Codes is used to ensure safe delivery of water. And untreated human waste and fecal matter are responsible for this which is why **Plumbing** holds so much importance. **"March 11"** is observed as "World **Plumbing Day**" globally as **Plumbing** plays such an important part in our day to day life's.

Why is *Plumbing* important?

Author SL Turner

The ability of *Plumbing* and sanitation systems to deliver clean water and remove waste has protected populations from communicable diseases. **Plumbing** advancements throughout history continue to protect lives in developing nations.

Which water pipe is best?

By far the most used pipe in **Residential Homes**, polyvinyl chloride (PVC) pipe is the *"white piping"* commonly used in plumbing applications. Affordable and versatile with several different fittings and sizes available, (PVC) is great for most warm and cold-water applications.

What are considered Plumbing Fixtures?

The most common **Plumbing Fixtures** are bathtubs, sinks, showers, tubs, toilets, and faucets. Pipes, drains and valves are part of a home's plumbing system that supply water to each of the **Plumbing Fixtures** and also drains wastewater away.

What are water conserving Plumbing Fixtures?

A non-complaint **Plumbing Fixture** is any of the following: Any toilet manufactured to use more than 1.6 gallons of water per flush, and any showerhead manufactured to have a flow capacity of more than 2.5 gallons of water per minute. Also, any interior faucet that emits more

Plumber 'Speaks'

than 2.2 gallons of water per minute. Plus, most urinals use 1 to 3 gallons of water per flush.

What is **Plumbing Network**?

Plumbing is the system or **network** of pipes, tanks, fittings, and other fixtures required to build water supply, heating storm and sanitation systems in a building. Heating, cooling, gas fitting, water removal, and supply of potable water systems are some of the important parts of this profession.

What is the difference between plumbing **Sanitary** and plumbing **Storm** systems?

The **sanitary sewer** is a system of underground **pipes** that carries sewage from bathrooms, sinks, kitchens, and other plumbing components to a wastewater treatment plant where it is filtered, treated and discharged. The **storm sewer** is a system designed to carry rainfall runoff and other drainage.

Do **Plumbers** work on gas lines?

Most people only consider using them for water pipe issues. However, specialty Plumbers can be used to work with natural gas lines, as well as other systems such as water sprinklers. Usually, a natural gas Plumber will have a license that states he or she can work on and install natural gas lines.

Author SL Turner

But today's plumbers to **not** work on gas lines that work mostly belongs to the steamfitters and local unions because of the size of the piping and because it's mostly welded pipe. That takes a certain expertise that most plumbers do not have especially because the weight in rigging of such big pipes and piping.

How do **Plumbers** check for **gas leaks**?

Using an air test and apply a soap-and-water solution to each connection in the gas lines. Look for bubbles to find leaks.

Plus, if you lose pressure in your gas piping. You now know you have a leak! Therefore, you must now find and fix the leaks.

Plumber 'Speaks'

Chapter #2

Lavs & Vanity Sinks

What does *"Lav"* mean in plumbing?

Lavatory "Lav" fixed bowl or basin with running water and drainage for washing. Many use the word "lavatory" meaning a bathroom.

What is a standard size bathroom sink?

There's no standard size for a bath sink. Some petite basins are just big enough for washing hands, while the largest sinks are big enough for washing hair or delicate clothing. Most round sinks are 16 to 20 inches in diameter, and most rectangular sinks are 19 to 24 inches wide and 16 to 23 inches front to back.

Bathroom Sink Styles:

Vessel Sink, Pedestal Sink, Console Sink, Wall-Mounted Sink, Vanity Sink.
Which is better ceramic or porcelain sink?

It is a ceramic-porcelain material that is, in appearance, much like cast iron enameled but with a stronger finish. This sink is fired, at double the temp than enameled sinks which gives them a nonporous surface and everlasting shine.

What is the best material to use for a bathroom sink?

Porcelain is extremely durable. There are many porcelains sinks are still in use. The porcelain material used to make sinks has glass and metal mixed in with the clay to give it extra strength and resistance to heat and chemicals.

Are porcelain sinks durable? What about S.S. sinks?

Stainless steel is resistant to heat and most chemicals, which is why they are also the choice of most laboratory sinks. A porcelain sink's durability is due to its construction.

Plumber 'Speaks'
Pedestal Sink

Single Bowl Sink
Stainless Steel

Is vitreous china better than ceramic?

Vitreous china is a ceramic material, such as porcelain, that is glazed with enamel. Vitreous china and porcelain cost about the same, however, vitreous china is better at resisting spills, scrapes, or other bathroom mishaps. The high gloss enamel is very durable and creates a stain-resistant surface.

Author SL Turner

Bathroom Sink
Wall Hung

Vanity Sinks

Plumber 'Speaks'

Chapter #3

Kitchen Sinks

What is a sink in kitchen?

Kitchen Sink also known by other names washbowl, hand basin and wash basin is a bowl-shaped plumbing fixture used for washing hands, dishwashing, and other purposes. Many sinks, especially in kitchens, are installed adjacent to or inside a counter.

Kitchen Sink "P" Trap
The curved pieces of drainpipe underneath your sink, commonly referred to as p-traps, do a lot of dirty work. Traps are made from ABS (black), PVC (white) or brass (either chrome-plated or natural colored). Traps come in 1 1/4 inch (standard bathroom sink) or 1 1/2 inch (standard kitchen sink) inside diameter sizes.

Author SL Turner

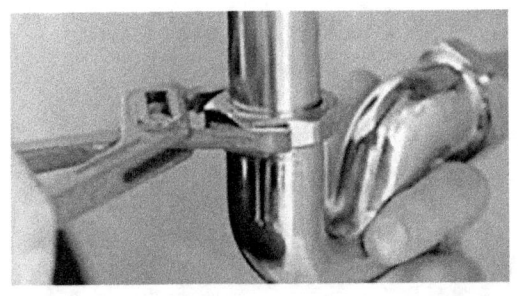

Double Bowl Kitchen Sink

What kind of kitchen sinks are in style?

Kitchen Sink Styles and Trends
Self-Rimming Sinks, Undermount Sinks:
Integrated Quartz Sink, Porcelain Apron Sink:
Integrated Marble Sink, Bamboo Apron Sink:
Prep Sinks, Iron Island Sink:

Plumber 'Speaks'
Double Bowl Kitchen Sink

Prep Utility Sinks:

Author SL Turner

Kitchen Sink

Single Bowl Sink

Plumber 'Speaks'

Chapter #4

<u>Toilets</u>

Which type of toilet is the most common and affordable type?

The most common **type of toilet** in the U.S. is the two-piece toilet.

Are Kohler or American Standard toilets better?

While **Kohler** is very comfortable and uses very little water per flush. When you compare these two toilet brands, it is evident that the **American Standard** is a better toilet due to its ease of use and efficiency in performance. **Kohler** is not far off as it has similar features and brings comfort.

What should I look for when buying a toilet?

Always buy a toilet with the correct rough-in. To determine the size of your toilet rough-in, measure from the wall behind the toilet to the bolt caps of your current toilet.

Most toilets are available in a 12-inch rough-in, which is the standard distance, but a 10- or 14-inch rough-in may be needed in some homes.

Which type of toilet is the most common and affordable type?

The most common **type of toilet** in the U.S. is the two-piece toilet.

Are Kohler or American Standard toilets better?

While **Kohler** is very comfortable and uses very little water per flush. When you compare these two toilet brands, it is evident that the **American Standard** is a better toilet due to its ease of use and efficiency in performance. **Kohler** is not far off as it has similar features and brings comfort.

What should I look for when buying a toilet?

Always buy a toilet with the correct rough-in.
To determine the size of your toilet rough-in, measure from the wall behind the toilet to the bolt caps of your current toilet.

Plumber 'Speaks'

While some places have western style **sit down/flush toilets**, the most common style of toilets in the Middle East is the squat toilet. A squat toilet is also known as a "Turkish Toilet". Squat toilets consist of a toilet bowel or pan that is at floor level.

What are the types of toilet?
According to the bowl design toilets can be classified around **four** types:

- Round bowl toilet.
- Square bowl toilet.
- Elongated Bowl toilet.
- Rectangular bowl toilet.

Most toilets are available in a 12-inch rough-in, which is the standard distance, but a 10- or 14-inch rough-in may be needed in some homes.

Question: Should a toilet closet flange fit into or over the PVC drainpipe?

They make both kinds. Just make sure it fits snug before you glue it. There are several types a 3x4 fits over 3" and inside 4" but there are ones that will go over 4" and inside 3" also.

Author SL Turner

Question: Why would a toilet bowl suddenly go dry?

A toilet like any fixture can lose its seal (water). By negative pressure or positive pressure or back siphonage. All are vent related except capillarity action.

The toilet is not vented properly and the action and turbulence of other plumbing in the building is causing the water to be sucked out of the bowl. There is usually a sink in the same bathroom, check to see if it has vent piping or is properly vented?

Two Piece Toilet

Plumber 'Speaks'

One Piece Toilet

Two Piece Toilet

Author SL Turner
Two Piece Toilet

Question: What is the best way to move the toilet flange a couple of feet and is there an adapter to go from cast iron to PVC?

Cut the toilet line down at the first 90 from the flange and relocate so that your toilet is still vented from its original vent. If it is dry vented, you need to move the toilet and the vent.

A <u>No-Hub Band</u> or <u>Fernco Coupling</u> is the preferred transition from cast iron to PVC plastic.

Question: What is the correct type seal for a toilet flange? Why should the toilet have this type seal?

Plumber 'Speaks'

Always use deep seal wax gasket. That type seals right inside and on top of the toilet flange, preventing sewer gases from entering the house.

Picture shows the correct placement of the plumbing deep seal wax gasket.

Note: Plumbers or handymen who don't seal toilet flange properly causes water leaks resulting in water damages, in thousands of dollars.

Question: What is the correct wax seal for a toilet flange that is flush with the floor?

Try Ultra Seal type wax gasket, I usually use the thickest wax ring I can find with the plastic funnel thing embedded in it.

Plumber 'Speaks'

Note: The below picture illustrates different parts of the toilet that the Plumber must check is working properly.

Please see the most important floor seal at the closet flange. Make sure to check and test that your toilet is fasten securely and does not rock. Did the Plumber seal your toilet fixture space and the floor?

Author SL Turner

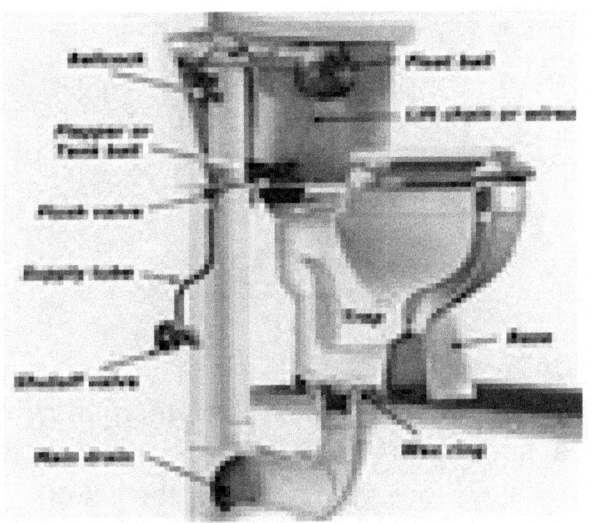

Plumber 'Speaks'

Chapter #5

<u>Water</u>

I Whenever you have a water emergency in the house such as a leak or water causing water damages to property. TURN-OF the water!

Question: How can you increase the water pressure of a faucet?

Usually the aerator is plugged; but, sometimes the rubber inside the stops under the sink wear out and tear away.

Sometimes the supply lines get kinked and need to be replaced. If you get a flex tube to connect to your stop under the sink, you could attach it and blow the water into a bucket to see if you have pressure there. If there is pressure, then you know the problem is above the stops.

Question: What causes a loud knocking sound like a jackhammer in the home plumbing when water is being shut off?

This is called "water hammer" and you need to install "water hammer arrestors" in the piping system where required.

Question: What is wrong when pipes make sound as you turn on the water?

It could be you have a loose washer on a faucet. If that is what you mean when turning on the water. Please install new washers and if the faucet become defective, then please buy and install new faucet.

Unscrupulous Contractors will say you need total new piping water replacements. Thanks for asking question, because just saved yourself plumbing bill worth many thousands of dollars.

Question: What causes or how to prevent water piping to rattle, shake and make loud noisy sounds? Does this mean major new water lines and new piping should be installed?

You should have Water Hammer Arrestors installed, same as the types and examples shown below. There are many Companies and Manufactures of Water Hammer Arrestors.

The water piping and water lines should be properly supported with the proper type of

Plumber 'Speaks'

insulation as required.

Question: How can I increase water pressure at my kitchen sink Faucet?

Author SL Turner

When you start having reduced water pressure at your kitchen sink or any sink. This means you may have some blockage in the aerator.

Clean the aerator out or replace with a new one. Then turn your water back on and your water pressure should have increased adequately.

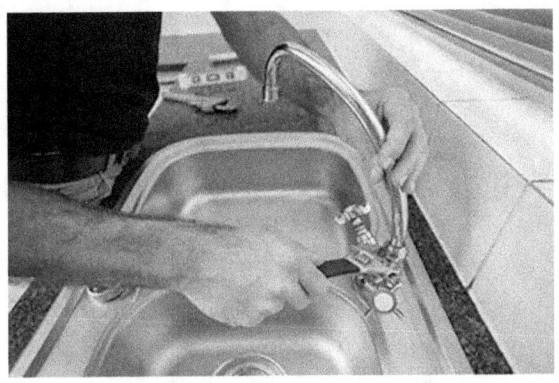

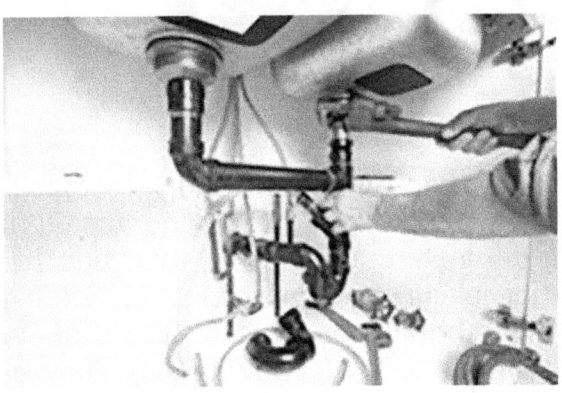

Chapter #6

Hot Water Tanks

Hot Water Tanks:

Hot Water Tank is a convenient **heat storage** medium because it has a high specific **heat** capacity. Water is non-toxic and low cost. An efficiently insulated tank can retain stored heat for days, reducing fuel costs. **Hot Water Tanks** may have a built-in gas or oil burner system, electric immersion heaters.

How do **Hot Water Tanks** work?

An **electric water heater** works essentially the same way as a **gas water heater**. It brings cold water in through the dip tube and heats it using the electric heating elements inside of the tank. The hot water rises in the tank and is moved throughout the home through the heat-out pipe.

Author SL Turner

What size **Hot Water Tank** do I need?

For 1 to 2 people: 30-40 gallons. For 2 to 3 people: 40-50 gallons. For 3 to 4 people: 50-60 gallons. For 5+ people: 60-80 gallons.

Hot Water Heaters:

Most demand **Hot Water Heaters** are rated for a variety of inlet temperatures. Typically, a 70°F (39°C) water temperature rise is possible at a flow rate of 5 gallons per minute through gas-fired demand water heaters and 2 gallons per minute through electric ones.

Faster flow rates or cooler inlet temperatures can sometimes reduce the water temperature at the most distant faucet. Some types of tankless water heaters are thermostatically controlled; they can vary their output temperature according to the water flow rate and inlet temperature.

What brand of **Hot Water Heater** is the most reliable?

What are the most reliable **Hot Water Heaters** on the market?

- Rinnai Water Heater. ...
- Ecosmart Tankless Water Heater.
- GE GeoSpring Water Heater. ...

Plumber 'Speaks'

- Stiebel Eltron Water Heater. ...
- Bosch Water Heater. ...
- Takagi. ...
- Kenmore Water Heater. ...
- American Standard Water Heater.
- Rheem

Are **Hot Water Heaters** too hot to touch?

Hot Water Heater is Hot to Touch. Even though it's designed to boil water, your water heater shouldn't get so hot that you can't touch it. Hot Water heaters are designed to be well insulated so that heat stays in.

How much does a **Plumber** charge to install a **Hot Water Heater**?

Plumbers typically charge $75-$150 per hour and can typically install a **Hot Water Heater** in a day (6-8 labor hours), for a total labor **cost** of $600-$840.

Why does the **hot water** run out?

The **sediment** buildup in your **Water Heater**.

This is the most likely reason your hot water is running out too quickly. It happens because water picks up natural minerals and sediment on its way to your Water Heater. Then, over time, the sediment sinks to the bottom of the Water Heater Tank because it's heavier than the water.

Author SL Turner

Can a **Water Heater** explode?

If the temperature is set too high or the pressure relief valve of a Water Heater malfunctions, a Water Heater can explode. This could happen with a gas or electric hot Water Heater. Although it is unlikely for Water Heaters to explode, but when they do, they operate much the same as a rocket.

What should my **Hot Water Tank** thermostat be set at?

A domestic hot water cylinder thermostat should be set at **60-65 C**. This is high enough to kill off harmful bacteria such as Legionella.

How do you know a **Hot Water Heater** is going bad?

Learn how to spot the signs of **Hot Water Heater** failure!

Lack of Hot Water. Popping or Rumbling Noises. Water heater noises. Cloudy Water. .

Leaking or Faulty Pressure Relief Valve.

Leaking Hot Water Tank.

Plumber 'Speaks'

Electric Hot Water Tank

Gas Hot Water Heater

Author SL Turner

Question: How do you stop a noisy hot water heater from making noise?

You can reduce the noise by draining the Water Heater and removing the lime deposits the best you can.

Most likely you can't reduce the noise if the gas Water Heater and you have hard water. The only way to fix this is to buy a new Water Heater and install a Water Softener.

Plumber 'Speaks'

Question: How do you stop pipes and water piping from sweating?

You need to wrap the pipes with insulation. Get insulation that the inside diameter of the insulation is larger than the outside diameter of the piping system.

Note: The wall thickness of the insulation should be at least **"one inch thick."** Make sure to use "duct tape" or "electrical tape" when covering turns in the piping system.

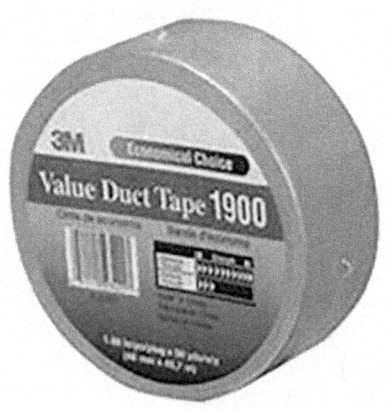

Author SL Turner

Question: Why would kitchen sink or washroom sinks have a stink smell?

The only reason I know that the stink smell would come about is that the P-Trap or U-Bend underneath a fixture type sink has lost its trap seal due to evaporation.

Refill the trap seal and pour water into the sinks to have a trap seal.

Question: What causes sink drains to make sucking sounds?

The sucking noise is a good thing, it means that the drain is working properly, and the noise comes from a swirl in the water going counterclockwise and pulls the water. The noise is only air.

Plumber 'Speaks'

Chapter #7
Sewers and Drains

Question: When roughing in a new toilet what is the correct distance from the center of the drain to the wall?

A minimum of **15** inches off the finished side wall and a minimum of **12** inches of the finished back wall.

However, **18"** side clearance is the international standard for wheelchair and assisted toilet maneuvering room. Maybe make more rooms comfortable for larger people.

Author SL Turner

Question: How do you make repairs when the Lavatory sink is plugged and not working? Plumber shown below is using the correct tool for the fixture trap removal. Once the fixture trap is removed and cleaned the Lav sink will drain properly.

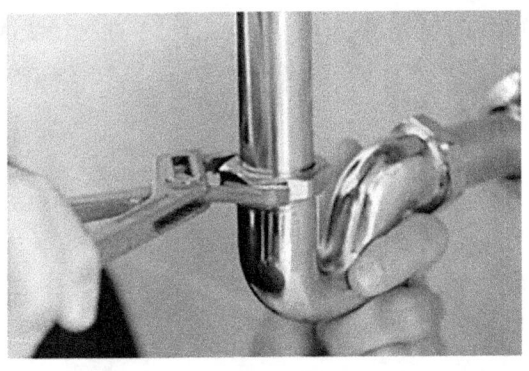

Question: What is waterproofing? Where should waterproofing be used?

Use waterproofing sealers and paints which are recommended materials for their usage only. However interior waterproofing sealers and paints will not alleviate all potential sources of basement leaking problem areas.

Question: How often should downspouts and gutters be cleaned?

Plumber 'Speaks'

Rain gutters and downspouts should be cleaned regularly. Remove all leaves, dirt and debris at least once or twice a year.

Question: What are "French Drains?" And where should they be used?

"French Drains" are used to channel water out of the building which then would be pumped out of the building through means of a sump pump and sump pump pit.

Question: How often should roofs be inspected?

Roofing systems should be inspected at least once a year.

Question: Who best installs new roofs? A General Contractor or a Roofing Contractor?

A licensed Roofing Contractor who offers a guarantee for their work and materials.

Question: How long does a roto-rooter job last?

It depends on what's wrong with your sewer. It could take anywhere from 10 minutes to a whole day. Usually blockages in the main occur either from roots or breakages in the pipe.

If the line has been snaked and you still have a problem, I would recommend having the sewer, camera for pictures of the interior sewer system piping.

Question: How do you keep roots out of the main sewer line?

The only thing you can really do is snake out the sewer drainage system and get it as clear as possible. Then you can dump copper sulfate down the system.

It probably won't kill all the roots, but it stops them from growing. This works best if you do it in the spring during the growing season. And you should probably do it about twice a year.

Another chemical to use is "Root X" because "Cooper Sulfate" is illegal in some states.

Putting any kind of chemicals down your sewer drain line is illegal in most countries around the world; it is considered an environmental hazard. If you are caught by authorities, you may be in serious trouble and the implications will be severe.

Plumber 'Speaks'

Therefore, it may be best to avoid using chemicals and rather rod or snake your sewer line, then cut down brushes or trees suspected to be growing over your sewer. Or replace your sewer line and maintain your sewer line by rodding it every six months.

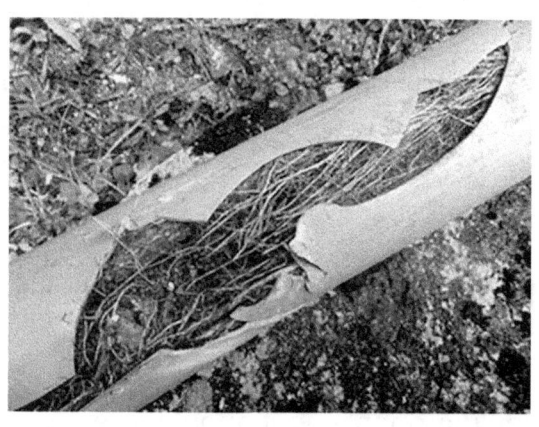

Question: What is a plumber's snake used for?

It is a tool they use to shove down your pipes to clean them out if they are clogged and the snake can be either electric powered or by hand. See the example sewer machines below:

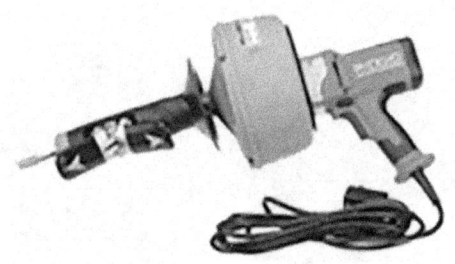

Chapter #8

Sump Pumps

Question: Every time it rains my basement gets flooded from water under basement floors and walls, what can I do about this situation and problem?

Seems like you may need a sump pump and basin used to manage surface runoff water and from underground water aquifers.

Author SL Turner

Question: How do you measure the water evaporation from your swimming pool?

Measuring Water Evaporation First; mark the water level on the wall of the pool. After some time, mark the water level again. It should be lower than the original one.

Measure the area of the pool and then multiply it by the difference of the two water levels. The result is the amount of water evaporated.

It is important to remember to turn off the pool's auto fill for this process. If you keep the auto fill on, then the pool level will not drop much, if any.

Question: What would cause water to seep out from under the toilet bowl onto the floor when the toilet is flushed?

You may have a partial clog somewhere in your pipes; therefore, the water is backing up in the first place.

Or the wax seal under the bottom of your toilet (you can't see it because it's UNDER the toilet) is broken.

Question: What causes the popping or tinkling sounds as hot metal pipes cool off?

Plumber 'Speaks'

Contraction, though the sounds are usually noisier when pipes are expanding as hot water runs through them. The sound you hear is caused by the expansion and contraction of the metal cause by the heating and cooling.

Question: What would cause the hot water to be rusty and brown?

Sometimes chemicals cause rust to get into the water lines. Or change in water pressure will cause this rust to turn loose and come through into the bathtub or other appliances.

If only the hot water is rusty! It's 100% your hot water tank!

The real reason why you're not seeing it in your cold water is because your cold-water lines run directly to your water fixtures.

Question: Who uses concrete roofs?

It is very common to find commercial and residential buildings having concrete roofing.

Author SL Turner

Chapter #9

Plumbing Answers

Faucets and drains are the parts of your plumbing system that are most likely to break down, so therefore you will need repairs more often. Faucets leak or drip and drains get clogged up. When you add these repairs to fixing toilet problems, you have covered almost everything you are likely to encounter.

Toilets: Clogged toilets are the most common plumbing problem can have. If a toilet overflows or flushes sluggishly clear the clog or backup with a plunger or closet auger. If the clog or backup persists, the problem may be in the mains waste and vent system piping.

A recurring puddle of water or water leak recurring on the floor or around the toilet it may be caused by a crack in the base of the toilet. This persistent puddle problem could possibly be

Plumber 'Speaks'

from the toilet tank. Also, check all water conditions to the toilet.

Toilet stability the toilet fixture shakes or rocks when it is not fastened securely. Check the closet flange and secure all loose connections including replacement of the closet wax ring gasket with nuts and bolts, too.

Water on the floor around toilet - check if the toilet base in the toilet tank is cracked and leaking. Sometimes you may have to insulate the toilet tank to prevent condensation. You may have to tighten the bolts and then check all water conditions.

Author SL Turner

Sink drains: every sink has a drain trap and a fixture drain. A sink gets clogs or plug ups by a buildup of soap and hair in the trap or fixture drain line. Remove clogs and plug ups by using a plunger, disconnecting and cleaning the trap or using a hand auger.

Clogged and plugged lavatory sinks can be cleared with a plunger. Remove the pop-up and strainer first, and then plug the overflow in the sink by stuffing a wet rag into it, allowing you to create air pressure with the plunger.

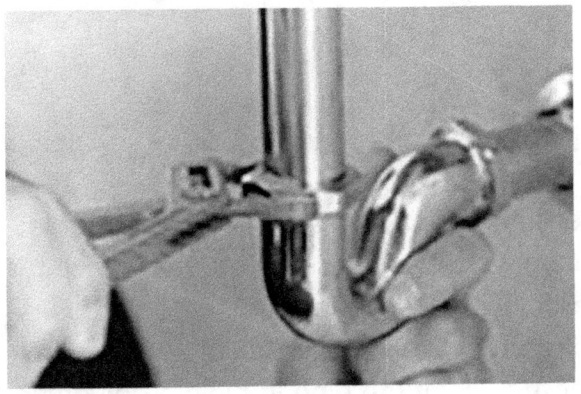

Dump out the debris after removing the trap. You may need a small wire brush when cleaning the trap bend. Reinstall the trap bend and tighten all

Faucets eventually just about all faucets develop leaks in drips. Repairs can be accomplished by

Plumber 'Speaks'

replacing the mechanical parts inside the faucet body. The main thing is to figure out what kind of faucet you have and know the make of its parts.

If your old faucet continues to leak after repairs are made, then you know that you will have to replace the old faucet with a new faucet.

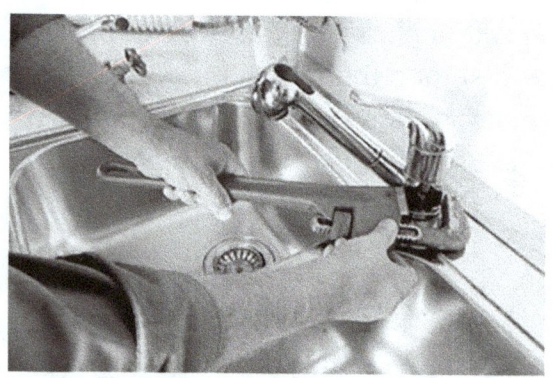

Water pressure at the spout of the faucet is problematic water pressure seems low water flow is partially blocked. Clean the faucet. This is this is not correct the situation, take a closer look under the sink if the pipe in his old galvanized piping, replace the corroded galvanized pipes with copper or plastic new water piping.

Sump pumps, price ranges from about $100-$500 or more, depending on the quality and the features of the sump pump. Below is a picture of a standard sump pump.

First decide between a pedestal and a submersible pump. The standard pump, non-submersible, as shown below:

Submersible pumps sit in the water a good deal of the time, they have a lifespan from 5 to 10 years. However, most manufacturers offer limited 1 to 5-year warranties. Sump pump is measured by horsepower.

It is better to buy a cast-iron sump pumps, which last longer than plastic or iron types because of corrosion. Make sure the power cord is long

Plumber 'Speaks'

enough because electrical extension cords are not to be used on sump pumps.

Always place pavers or bricks underneath the sump pump so that mud, dirt, and grit not plug up the operation of the pump.

Using a plunger or a "plumber's plunger" which is about 10 to 12 feet long, to push the obstruction down thru F.A.I. and hose trap, to clear the clogged or plugged up sewer line.

For some hot water down the line and then apply the pressure forced by plunging several times.

Author SL Turner

The pressure from the simple tool can generate pressure to blow out obstructions quite quickly.

Using an **electric snake** if the obstruction is in the form of tree roots, instead of common obstructions like toilet paper rags leavings, and garbage. You need to use a snake tool, which is normally used by plumbers and available in the hardware stores, for a rental fee.

In cases of extreme clogging and plug ups, call a professional plumbing company may use a jet hydro-vac machine.

A clogged sewer is a problem, which needs immediate attention as it can become a nasty nightmare if not taking care of in time.

The **below** is the type electric sewer machine or **electric snake machine** that you will need:

Plumber 'Speaks'

Author SL Turner

Hot Water Heater problems normally become self-evident. The hot water faucet fails to summon or get hot water from the spout, you see scribbling or puddles near the Hot Water Heater, or the tank emits strange gurgling or pop and sounds coming from the Hot Water Heater.

Hot Water Heaters makes strange noises expanding and contracting metal parts, or more likely, minerals and higher water scale accumulations inside the tank can cause the noises coming from the Hot Water Heater.

To avoid scale buildups, every few months, open the **drain valve** at the base of the tank, and flush the tank until the rest runs out and you see clearwater.

Most Hot Water Heaters manufacturers recommend draining and flushing your hot water tank once per year or every six months and heart water areas. This helps remove sediment and minerals that collect at the bottom of the tank. Sometimes the sediments chunks may be too large to pass through the drain valve on the tank.

If you hear a boiling sound of water coming from inside your tank this could indicate overheating and very dangerous pressure buildup. You should call a plumber or service professional immediately.

Patch burst pipes, if the water pipe freezes and breaks, your priority may be getting it working again, at whatever it takes get the job done. Just remember you are getting a temporary patch job, you still need the pipe or pipes fixed permanently.

Kitchen sinks come in many stylish designs now, but you can get basic stainless-steel sinks for around $100-$200. A higher quality stainless steel sink has a higher thicker gauge steel in a higher amount of nickel alloy and surface finish.

Author SL Turner

The high-quality stainless-steel sink should stay bright for many years and higher quality may cost from $200-$400.

Cast-iron sinks, with enamel finish are very popular. It is possible to chip them, but they have very strong and very durable. Most cast-iron sinks are self-rimming, meaning they have raised lips they rest on top of the countertop; some are available with flush fit rims.

Plumber 'Speaks'

Types of shower almost nothing is more invigorating than the nice shower. If you have a good showerhead that you use. The type of showerhead you use will have a huge impact on the quality of your shower. There have been many advances in plumbing fixtures, so that your shower heated options are virtually unlimited.

Think showerhead is are fixed directly on the shower wall cannot be removed. However, some fixed chalets are adjustable and can be moved or aimed in different directions, which can be convenient when tall and short people live in the same house share the same shower.

Handheld showerheads are connected to a flexible hose that is mounted on the shower wall. With handheld showerheads, you can remove the shower. The attached hose usually allows for greater range of motion, making this the perfect showerhead for bathing pets, clean in the shower stall and hand washing your clothes.

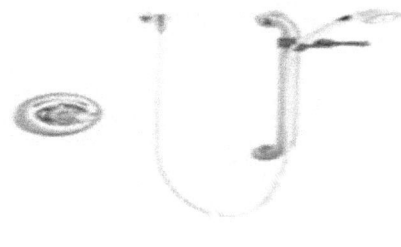

Author SL Turner

Low-flow shower heads many consumers are turning to low-flow showerheads to reduce energy costs. These showerheads can be fixed or handheld and greatly reduce the amount of water that is sprayed from the nozzle.

The bathtub uses something called a "trip lever" as the waste system on the tub. There is also a P-trap connected to the trip lever assembly. Most of the time the P-trap will not be clogged or plugged up.

The drain will just be full of hair and can be cleaned out in several ways. How much hair and debris that is wrapped around the tub trip lever. Once you have cleaned all the visible here and debris on the trip lever and drain, apply parts with a light coating of grease.

Take a small hand snake or small electric snake and run 2 or 3 feet of cable to make sure that there are no obstructions in the drain line.

Plumber 'Speaks'

Author SL Turner

Chapter #10

Water, PEX - Piping

What are pipe adapters?

Adapters are an extremely important pipe fitting that extend or terminate pipe runs. They are used to connect dissimilar pipes. These fittings are somewhat similar or like pipe couplings, with the difference that they connect pipe of different types, one of which is an IPS (Iron Pipe Size).

What are pipe fittings used for?

Pipe fittings, also known as pipe connectors, attach one pipe to another in order to lengthen the run or change the flow direction in a plumbing system. They are used to combine, divert or change the flow of the water supply, and they come in a variety of sizes to fit the pipe they will connect.

Plumber 'Speaks'

What are the different types of pipe fittings?

Pipe Fittings used in piping work are mainly Elbow, Tee, Reducer, Union, Coupling, Cross, Cap, Swage Nipple, Plug, Bush, Joint Adapters, Olet (Weldolet, Sockolet, Thredolet, Nipolet) Flanges and Valve.

What is the difference between coupling and adapter?

If the two ends of a coupling are different (e.g. one BSP threaded and one NPT threaded), then it is usually referred to as an adapter.

Another variation would be 3/4" NPT to 1/2" NPT. When the two ends use the same connection method but are of a different size, the terms reducing coupling or reducer are used.

What is PVC pipe?

PVC Pipe, polyvinyl chloride, pipe is a drain or vent line type of plumbing pipe. PVC initially gained popularity because it was lighter and easier to work with than traditional galvanized steel pipe.

Author SL Turner

How does a pipe coupling work?

Coupling is a solid fitting with female threads on the inside of both ends. It is used for joining two lengths of pipe together that are not locked in place and can be turned.

A union is used when you are trying to join two pipes together that are fixed, thus unable to be turned.

How do you connect PVC pipe to fittings?

Push the pipe into the PVC pipe connectors fitting and twist until the marks are aligned. Press and hold the pipe for about 15 seconds.

Just swiping the pipe with cement and pushing on the fitting won't ensure a strong joint. You want to make sure you have an even layer of cement over, all the surfaces fitting and pipe.

PEX Piping for Potable Water Systems.

Available in three colors (red, blue, and white) **PEX tubing** is ideal for a wide range of residential and commercial plumbing applications. It is approved for all potable water systems, as well as for use in crimp, clamp, or

Plumber 'Speaks'

push-fit systems. It is easy to use, easy to install, and can offer you and your clients a wide range of benefits over traditional pipe installations.

PEX Piping for Potable Water Systems.
Red, Blue, and White Tubing

PEX Piping for Potable Water Systems.

Listed to NSF 61, NSF 14 & CSA B137.5 for use in drinking water systems.

Can be used with crimp, clamp, or SharkBite® push-connect fittings.

Flexible tubing can be installed almost anywhere you need - suitable for direct bury or behind-the-wall installations.

Author SL Turner

Affordable materials and fewer fittings required plus fast installation reduces overall costs without sacrificing reliability and performance.

Resists corrosion and scale buildup.

Does not cause water hammer when supply is shut off.

Not as thermally conductive as some other types of pipe, so less heat is lost while water is in the pipe, saving you energy and water

PEX Piping for Potable Water Systems.

What is the difference between the different colors (red, blue, white) of PEX pipe? Do I have to use red for hot and blue for cold?

There is no difference between the different colors of PEX except the color. They are all made of the same material in the same manner and designed for the same applications.

You do not have to use red for hot and blue for cold. You can use any color of PEX tubing you want for any potable water line you want.

Plumber 'Speaks'

Some people just find it easier to work with the different colors to help them keep track of hot vs. cold water lines.

And, some people prefer to use white for both since it looks more like the rigid plastic piping they're used to.

PEX Piping for Potable Water Systems.

PEX crimp clamp rings
Clamp and rubber ring

Author SL Turner

PEX (ball valve)

PEX (valve)

PEX – compression angle valve

Plumber 'Speaks'

PEX – coupling

PEX – 90-degree elbows

PEX – plug

Author SL Turner

PEX – reducing coupling

PEX – tee

PEX – reducing elbow

Plumber 'Speaks'

PEX – sweat female adapter

PEX – sweat male adapter

PEX – threated male adapter

Author SL Turner

PEX – swivel adapter

Are the clamp style or crimp style rings better?"

Each system has advantages and disadvantages in certain applications. **The stainless-steel clamp rings** are naturally non-corrosive, which may make them a better fit for coastal areas and are typically easier to install in confined spaces.

The crimp rings are more suitable for use with polybutylene or plastic fittings since copper is a softer metal and will provide a better seal and can be a better fit for areas where expansion is a part of life.

Some may think the fact that the clamps are easier to remove is a bonus, while others worry about the potential for leaks with such an easy to remove fitting.

Plumber 'Speaks'

Successful clamp connections can be confirmed based on the release of the ratchet tool, crimp fittings can be checked with a go-no-go gauge long after the original connection was made, which is reassuring. Really, the choice boils down to personal preference.

PEX Piping for Potable Water Systems.

Can I use the PEX Clamp Ring Tool on copper crimp rings?

PEX Clamp Ring Tool is for clamping the stainless-steel clamp rings on to PEX fittings and pipe. If you are using copper crimp rings, then you must use one of the **Crimp Ring Tools**. The Clamp Ring Tools and the Crimp Ring Tools are not *interchangeable.*

Is there a removal tool for stainless steel clamp rings?

No, a special tool is not needed for the removal of a stainless-steel clamp ring. Just use a flat head screwdriver to pop the clamp on the side.

Author SL Turner

Is PEX plumbing any good?

PEX is an excellent piping material for hot and cold-water systems, especially since it is flexible and well adapted for temperatures below freezing all the way up to 200 degrees Fahrenheit. PEX is easy for plumbers to install and has fewer joints, bringing your costs down.

What is PEX pipe made of?

Almost all PEX is used for pipe and tubing is made from high-density polyethylene (HDPE). PEX contains cross-linked bonds in the polymer structure, changing the thermoplastic to a thermoset. Cross-linking is accomplished during or after the extrusion of the tubing.

Can PEX pipes freeze and burst?

Due to its flexibility (polymers), PEX has a small margin of expansion under the damaging pressure caused by ice formation. If the weather gets cold enough, PEX pipes can and will freeze like any other plumbing pipe.

Plumber 'Speaks'

Why is PEX plumbing bad?

It is strictly designed for indoor and underground use as water piping. It's resistant to freezing temperatures but will not resist a direct and hard freeze. It's also not good for PEX to be exposed to direct sunlight outside for long periods as it will break down the material's strength and lead to full failure.

How long will PEX last?

PEX piping is more durable than copper piping, and it is used in an innumerable number of applications, such as radiant heating systems for residential and industrial plumbing. The average life span of PEX for plumbing purposes goes well over 50 years.

What is better copper or PEX?

PEX doesn't degrade like copper, higher PSI rating freezing pipes will still burst, but PEX will be able to handle more freezing water
than copper. PEX Tubing is much more resistant to freeze-breakage than copper or rigid plastic pipe. PEX Tubing is cheaper because it takes much less labor to install.

Author SL Turner

Should I replace copper with PEX?

If it's a small area that's leaking, you could just replace that section with either PEX or copper. PEX costs less than copper. Coupled with the quicker installation, the savings over installing copper pipes can be significant. Also, PEX pipes don't corrode like copper.

Plumber 'Speaks'

Cost Estimating

Author SL Turner

Chapter #11

Cost Estimating

The Author, Master / Plumber has written this book specifically benefit **Homeowners** and **Small Businesses** who face those tough decisions trying to save money, because upkeep and repairs are costly.

This book takes you by the hand and helps explain the ***do's*** and ***don'ts***. Most plumbing books say *do this* but never say *don't do that*. Confusing isn't it!

The most important reason why I wrote this book to help fight against unscrupulous **Contractors**. Who sometimes are always ripping-off the elderly and the disabled!

This book gives **Homeowners** and **Small Businesses** an opportunity to evaluate Contractors prices and scope of work.

Plumber 'Speaks'
Cost Estimating Jobs

When designing a building or developing software, successful projects require accurate cost estimates. Cost estimations forecast the resources and associated costs needed to execute a job or project. This helps you achieve project objectives within the approved timeline and budget.

Cost estimating is a well-developed discipline. By understanding cost estimating and using standard estimation techniques, you can improve your forecasts. The book (Home Guide) is complete guide to project cost estimating! Computer charts will walk you through the key concepts and major estimating techniques.

Today everyone has a **Computer**. To track and estimate costs, you need estimating software program. As an expert estimate of 40 years I have developed computed labor and cost Charts to easily set-up in your **Computers**.

Using my free Charts, labor rates equal hours times the labor hourly costs, gives the estimated costs. I paid over $10K for the software. I am too happy to give and save **Homeowners** and **Small Businesses** savings of $10K or more.

Author SL Turner

What are the four categories of estimating software?

For **Construction Project Development** and control, there are **four** basic types of cost estimates that are developed and used by DOE and its contractors. These estimates are planning, feasibility, study, estimates, budget or design estimates.

Students can easily use free Charts, labor rates equal hours times the labor hourly costs, gives the estimated costs.

Plumber 'Speaks'

This book helps guarantee to save you thousands of dollars and helps prevent **Contractor** rip-offs, too!

What is a **Direct** cost?

Answer: **Direct cost** includes materials, labor, expense, or distribution **cost** associated with producing a product. It can be easily traced to a product, department or project.

What is **Cost Estimating**?

A **cost estimate** is the approximation of the cost of a program, project, or operation. The **cost estimate** is the product of the **cost estimating** process. The **cost estimate** has a single total value and may have identifiable component values.

What are the three basic types of **Cost Estimating**?

Nonetheless, there are three types of cost estimation classified according to their scope and accuracy. These are (**1**) order of magnitude estimate; (**2**) budget estimate; and (**3**) definitive estimate.

What is a **Budget** cost estimate?

Answer: budget estimates are generally prepared to form the basis for budget authorization, appropriation, and/or funding. As such, they typically form the initial control estimate against which all actual costs and resources will be monitored

What is the purpose of Estimating and Costing?

Estimate and **Costing** serves several purposes in the construction process including preparation and finalize of bids and cost control. The main purpose is to provide volume of work for cost control and to see that the adequate options of materials are explored during the execution of the project.

What is an estimate in project management?

Cost estimation in project management is the process of forecasting the financial and other resources needed to complete a project within a defined scope. Once the project is in motion, the cost estimate is used to manage all of its affiliated costs in order to keep the project on budget.

What is Project cost estimate?

Plumber 'Speaks'

Project cost estimation is the process of predicting the quantity, cost, and price of the resources required by the scope of a project.

Since cost estimation is about the prediction of costs rather than counting the actual cost, a certain degree of uncertainty is involved.

What is ballpark estimate in project management?

The **Ballpark Estimate** is also known as the Rough Order estimate is based on high-level objectives, provides a bird's-eye view of the project deliverables, and has lots of wiggle room.

Why is cost estimate important in project management?

Cost estimate helps in identifying project risks and provides an option to discuss alternatives.
An accurate project costing plan enables one to weigh anticipated benefits against anticipated costs to see if the project makes sense.

Author SL Turner

Estimating Piping & Fittings Labor

<u>Notice:</u> All labor rates are reference as average, depending on different job material applications. Labor rates should be adjusted accordingly.

Labor = parts of an hour = .10 = 6 minutes
Labor = parts of an hour = .25 = 15 minutes
Labor = parts of an hour = .5 = 30 minutes
Labor = parts of an hour = .75 = 45 minutes
Labor = parts of an hour = 1.0 = 60 minutes

DWV Copper Pipe Size	Pipe Labor	Fitting Labor
1 1/2	0.01	0.1
2	0.01	0.1
3	0.03	0.15
4	0.04	0.25
6	0.06	0.5

Plumber 'Speaks'

Estimating Piping & Fittings Labor

<u>Notice:</u> All labor rates are reference as average, depending on different job material applications. Labor rates should be adjusted accordingly.

Labor = parts of an hour = .10 = 6 minutes
Labor = parts of an hour = .25 = 15 minutes
Labor = parts of an hour = .5 = 30 minutes
Labor = parts of an hour = .75 = 45 minutes
Labor = parts of an hour = 1.0 = 60 minutes

Copper "L" Pipe Size	Pipe Labor	Fitting Labor
1/2	0.01	0.1
3/4	0.01	0.1
1	0.01	0.1
1 1/4	0.02	0.15
1 1/2	0.02	0.15
2	0.02	0.25
3	0.03	0.35
4	0.04	0.5

Author SL Turner

Estimating Piping & Fittings Labor

<u>Notice</u>: All labor rates are reference as average, depending on different job material applications. Labor rates should be adjusted accordingly.

Labor = parts of an hour = .10 = 6 minutes
Labor = parts of an hour = .25 = 15 minutes
Labor = parts of an hour = .5 = 30 minutes
Labor = parts of an hour = .75 = 45 minutes
Labor = parts of an hour = 1.0 = 60 minutes

PVC #40 Pipe Size	Pipe Labor	Fitting Labor
1/2	0	0.03
3/4	0	0.03
1	0	0.03
1 1/4	0.01	0.03
1 1/2	0.01	0.03
2	0.02	0.03
3	0.03	0.05
4	0.03	0.05

Plumber 'Speaks'

Estimating Piping & Fittings Labor

<u>Notice:</u> All labor rates are reference as average, depending on different job material applications. Labor rates should be adjusted accordingly.

Labor = parts of an hour = .10 = 6 minutes
Labor = parts of an hour = .25 = 15 minutes
Labor = parts of an hour = .5 = 30 minutes
Labor = parts of an hour = .75 = 45 minutes
Labor = parts of an hour = 1.0 = 60 minutes

Black Steel Pipe Size	Pipe Labor	Fitting Labor
1/2	0.01	0.1
3/4	0.01	0.1
1	0.02	0.1
1 1/4	0.02	0.15
1 1/2	0.02	0.15
2	0.03	0.25
3	0.04	0.35
4	0.06	0.5

Author SL Turner

Estimating Piping & Fittings Labor

<u>Notice:</u> All labor rates are reference as average, depending on different job material applications. Labor rates should be adjusted accordingly.

Labor = parts of an hour = .10 = 6 minutes
Labor = parts of an hour = .25 = 15 minutes
Labor = parts of an hour = .5 = 30 minutes
Labor = parts of an hour = .75 = 45 minutes
Labor = parts of an hour = 1.0 = 60 minutes

No-Hub CI Pipe Size	Pipe Labor	Fitting Labor
1 1/2	0.01	0.1
2	0.01	0.1
3	0.02	0.15
4	0.03	0.25
6	0.04	0.35
8	0.06	0.5
10	0.08	0.75
12	0.1	1

Plumber 'Speaks'

Estimating Piping & Fittings Labor

<u>Notice:</u> All labor rates are reference as average, depending on different job material applications. Labor rates should be adjusted accordingly.

Labor = parts of an hour = .10 = 6 minutes
Labor = parts of an hour = .25 = 15 minutes
Labor = parts of an hour = .5 = 30 minutes
Labor = parts of an hour = .75 = 45 minutes
Labor = parts of an hour = 1.0 = 60 minutes

Galv. Steel Support Size	Pipe Labor	Fitting Labor
1 1/2	0.01	0.1
2	0.01	0.1
3	0.02	0.15
4	0.03	0.15
6	0.04	0.25
8	0.06	0.25
10	0.08	0.35
12	0.1	0.35

Author SL Turner

Estimating Piping & Fittings Labor

<u>Notice:</u> All labor rates are reference as average, depending on different job material applications. Labor rates should be adjusted accordingly.

Labor = parts of an hour = .10 = 6 minutes
Labor = parts of an hour = .25 = 15 minutes
Labor = parts of an hour = .5 = 30 minutes
Labor = parts of an hour = .75 = 45 minutes
Labor = parts of an hour = 1.0 = 60 minutes

Copper Type Support Size	Pipe Labor	Fitting Labor
1 1/2	0.01	0.1
2	0.01	0.1
3	0.02	0.15
4	0.03	0.15
6	0.04	0.25
8	0.06	0.25
10	0.08	0.35
12	0.1	0.35

Plumber 'Speaks'

Estimating Piping & Fittings Labor

<u>Notice:</u> All labor rates are reference as average, depending on different job material applications. Labor rates should be adjusted accordingly.

Labor = parts of an hour = .10 = 6 minutes
Labor = parts of an hour = .25 = 15 minutes
Labor = parts of an hour = .5 = 30 minutes
Labor = parts of an hour = .75 = 45 minutes
Labor = parts of an hour = 1.0 = 60 minutes

Steel Hangers Pipe Size	Pipe Labor	Fitting Labor
1/2	0	0.1
3/4	0	0.1
1	0	0.1
1 1/4	0.01	0.15
1 1/2	0.01	0.15
2	0.02	0.15
3	0.03	0.25
4	0.03	0.25

Author SL Turner

Estimating Piping & Fittings Labor

<u>Notice</u>: All labor rates are reference as average, depending on different job material applications. Labor rates should be adjusted accordingly.

Labor = parts of an hour = .10 = 6 minutes
Labor = parts of an hour = .25 = 15 minutes
Labor = parts of an hour = .5 = 30 minutes
Labor = parts of an hour = .75 = 45 minutes
Labor = parts of an hour = 1.0 = 60 minutes

Cop. Hangers Pipe Size	Pipe Labor	Fitting Labor
1/2	0	0.1
3/4	0	0.1
1	0	0.1
1 1/4	0.01	0.15
1 1/2	0.01	0.15
2	0.02	0.15
3	0.03	0.25
4	0.03	0.25

Plumber 'Speaks'

Estimating Piping & Fittings Labor

<u>Notice:</u> All labor rates are reference as average, depending on different job material applications. Labor rates should be adjusted accordingly.

Labor = parts of an hour = .10 = 6 minutes
Labor = parts of an hour = .25 = 15 minutes
Labor = parts of an hour = .5 = 30 minutes
Labor = parts of an hour = .75 = 45 minutes
Labor = parts of an hour = 1.0 = 60 minutes

Split Ring Galv. Steel	Pipe Labor	Fitting Labor
1/2	0	0.05
3/4	0	0.05
1	0	0.05
1 1/4	0.01	0.1
1 1/2	0.01	0.1
2	0.02	0.1
3	0.03	0.15
4	0.03	0.15

Author SL Turner

Estimating Piping & Fittings Labor

<u>Notice:</u> All labor rates are reference as average, depending on different job material applications. Labor rates should be adjusted accordingly.

Labor = parts of an hour = .10 =   6 minutes
Labor = parts of an hour = .25 = 15 minutes
Labor = parts of an hour = .5 = 30 minutes
Labor = parts of an hour = .75 = 45 minutes
Labor = parts of an hour = 1.0 = 60 minutes

Split Ring Copper	Pipe Labor	Fitting Labor
1/2	0	0.05
3/4	0	0.05
1	0	0.05
1 1/4	0.01	0.1
1 1/2	0.01	0.1
2	0.02	0.1
3	0.03	0.15
4	0.03	0.15

Plumber 'Speaks'

JOB NO.#	
Fixtures and Equipment **Description**	**Labor hours**
Water Closet (toilet)	1.5
Lav (hand-sink)	1.5
Bath-Tub and Shower	8
Shower	2
Kitchen Sink (single bowl)	1.5
DishWasher	1.5
Icemaker	1
Hot Water Tank	2
Vanity Sink	1.5
Kitchen Sink (double bowl)	2
Sump Pump	0.5

About the Author

Master / Plumber, Sherman Turner has work as a Plumbing Inspector, a Journeyman Plumber, and a Plumbing Estimator. As a businessman he uses all three (3) levels of his experience to serve the people in the communities.

Sherman Turner is a *Master / Plumber* and an Author. He has written other helpful books on plumbing and here is list of many books you can search on Amazon.com for the below listed Plumbing and Estimating books.

Newly Released books are (The Plumber) (Plumbing Answers 101 & Tips) and (Plumbing Estimating 101). On *Amazon*.com.

Plumber 'Speaks'

www.ingramcontent.com/pod-product-compliance
Lightning Source LLC
Chambersburg PA
CBHW070806220526
45466CB00002B/561